SCOTT M. GRAFFIUS

agile TRANSFORMATION

A Brief Story of
How an Entertainment Company
Developed New Capabilities
and Unlocked Business Agility
to Thrive in an Era of Rapid Change

(This page is intentionally blank.)

(This page is intentionally blank.)

Table of Contents

"Brief but excellent book on how to perform an agile transformation"

— Review on Amazon

"I read 1-2 books on agile a month and *Agile Transformation* is one of the best ones I've read in years. It tells a story from start to finish about a real Scrum implementation. I found it practical and insightful. It gave me lots of ideas about working with different people/groups (stakeholders/sponsors, scrum master, product owner, development team) more effectively to better achieve agility and deliver business value. I can't recommend this enough!"

— Review on Goodreads

Praise for *Agile Transformation*

"Scott M. Graffius has another winner with his second book, *Agile Transformation*." His first book was a multi-award-winning bestseller on agile Scrum, and now he has written a wonderful follow-up detailing the story of the genesis of his first book idea. *Agile Transformation* is a succinct, yet well-written, first-person account of Scott's journey with learning the tenets of agile scrum, in the context of creating adoption and assimilation at an entertainment company. Anyone who is interested in learning about agile Scrum (or even navigating enterprise-level technology adoption in general) would find this book highly informative and of great interest."

— Review on Amazon

Agile Transformation:
A Brief Story of How an Entertainment Company Developed New Capabilities and Unlocked Business Agility to Thrive in an Era of Rapid Change

Scott M. Graffius

This work is based on a true story. Identifying details have been changed.

Book

Graffius, Scott M. (2019). *Agile Transformation: A Brief Story of How an Entertainment Company Developed New Capabilities and Unlocked Business Agility to Thrive in an Era of Rapid Change*. North Charleston, SC: CreateSpace.

ISBN-10: 1072447967
ISBN-13: 9781072447962

Publisher: CreateSpace, 4900 LaCross Road, North Charleston, SC 29406

eBook

Graffius, Scott M. (2019). *Agile Transformation: A Brief Story of How an Entertainment Company Developed New Capabilities and Unlocked Business Agility to Thrive in an Era of Rapid Change*. Seattle, WA: Amazon Digital Services.

ASIN: B01FZ0JIIY

DOI: 10.5281/zenodo.2652976

Publisher: Amazon Digital Services, 410 Terry Avenue North, Seattle, WA 98109

The storyline was originally published by the author in a different form in 2019. This publication includes an updated version of the story and additional content.

Version 19042503.19061103

Introduction

Thriving in today's marketplace frequently depends on making a transformation to become more agile. Those successful in the transition enjoy faster delivery speed and ROI, higher satisfaction, continuous improvement, and additional benefits.

As a Principal Consultant and the CEO of Exceptional PPM and PMO Solutions™, I help companies strengthen their project management capabilities and realize their strategic objectives and business initiatives. A fantastic agile transformation experience and result with a client organization in the entertainment industry is the subject of this brief story. The agile journey was also the inspiration for my first book, *Agile Scrum: Your Quick Start Guide with Step-by-Step Instructions*. The publication garnered 17 first place awards from national and international competitions, and BookAuthority named it "One of the Best Scrum Books of All Time."

In *Agile Transformation: A Brief Story of How an Entertainment Company Developed New Capabilities and Unlocked Business Agility to Thrive in an Era of Rapid Change,* I share a behind-the-scenes account of a successful agile implementation at a global entertainment company. The story is based on actual events and it's told from my perspective as an agile coach. The transformation dramatically improved the way the organization works and delivers business value. New capabilities and practices enabled the enterprise to adapt to its changing environment, move faster, and drive innovation, which made it more competitive and prosperous.

Agile Transformation is based on a true story. Identifying details have been changed. The glossary—located at the back of the book—explains terms used in the story. I hope you enjoy this book.

The Transformation

Part One

The Call

The assistant to the executive vice president (EVP) of technology for a division of a global entertainment company contacted me by phone. I was told that I was referred by someone who knew me and thought I would be a good fit for contract work. A one-hour consultation with the EVP was scheduled for the next day. I then received an email with several attachments including a non-disclosure agreement which needed to be executed and brought to the meeting.

I brought the documents to the session. The EVP explained that his division of the company was experiencing an alarming trend of problems with project management. He reported that deliverables were not meeting expectations, there was a marked decline in satisfaction, and—this was characterized as "the straw that broke the camel's back"—a highly skilled and very well-respected team member quit, citing the problems as her reason for departure.

After an hour with the EVP, we agreed to extend the meeting—my complimentary consult—an additional hour into billable time. I learned that this division of the company previously used a traditional plan-driven/ waterfall approach for development and delivery. The EVP explained that things worked reasonably well then and that they were "close enough" to being on-scope, on-budget, and on-time that team members and stakeholders were satisfied.

The EVP explained, however, that he wanted to adopt a model (Scrum) employed by some of the other divisions of the company. The EVP said that he conducted a search for a value-added reseller (VAR) to help his group move to agile. A VAR-partner of a popular software solution was selected, and the VAR transitioned the group to Scrum. Reportedly, very few problems surfaced during the VAR's contractual engagement, which ended two weeks into the first sprint (then, the duration for sprints was four weeks). Many problems surfaced subsequently, however. The VAR had been gone for two months at the time of my meeting with the EVP.

The EVP asked me to resolve the problems by re-implementing agile. I explained that change depends on many people and multiple factors and that a specific end result could not be guaranteed. I suggested, however, that I'd be honored to work with his group in the first step—an assessment—which would help inform the subsequent work of foundational planning. The objectives of the assessment include working closely with the EVP, the Scrum Team and stakeholders to understand their goals—and the environment, roles and practices. He asked me when I could start. I asked him when he needed me, and he replied "immediately." I agreed to start the next day.

Part Two

The Goals

Highlights related to the goals follow.

The EVP identified his top three desired outcomes:

1. Accelerate the development and delivery of products and services—to be faster than the earlier baseline of 6-12 months,
2. Improve the satisfaction of the Scrum Team—to be the same or better than it was earlier with the traditional/waterfall approach, and
3. Improve the satisfaction of stakeholders—to also to be the same or better than it was earlier with the traditional/waterfall approach.

I individually met with and carefully listened to each member of the Scrum Team. We then met as a group, and I asked them to identify their top one, two or three goals. They discussed the matter, voted, and decided on these:

1. Meet or exceed the expectations of management, and
2. Deliver valuable products.

I individually met with executives from different departments (the stakeholders). In every case, they related that things were OK with the earlier traditional/waterfall approach, but that things were worse now. The stakeholders indicated that, with the earlier approach, someone on the project team worked with their group to gather requirements, and around 6-12 months later the results were deployed. However, it was reported that with the then-current approach, the projects' purpose and requirements were not understood and that what was produced was unusable. The goals of the stakeholders were:

1. For them or their representatives to be as—or more—involved as earlier with the traditional/waterfall approach, and
2. Get a usable product as often as—or more frequently than—earlier with the traditional/waterfall approach.

Here's a recap of everyone's goals. The EVP wants the development and delivery of products and services to be faster than 6-12 months, improved satisfaction of the Scrum Team, and improved satisfaction of stakeholders. The Scrum Team wants to meet or exceed expectations of management and deliver valuable products. The stakeholders want (themselves or via their representatives) to be more involved in requirements/ user stories and get useable project-delivered products more frequently than every 6-12 months.

Everyone permitted their goals to be shared with others. After discussing the subject with the EVP, I wrote the goals on oversize paper and posted it in a common area proximate to the Scrum Team and accessible to the stakeholders.

Part Three

The Environment

Highlights on the environment including roles and practices—primarily focused on the Scrum Team—follow.

I learned that the Scrum Team was composed of 16 people: one Scrum Master, one Product Owner, and 14 Development Team members. Both the Scrum Master and the Product Owner explained that they were familiar with agile, but that they had no prior work experience with agile/Scrum and no related training—except for what was provided by the VAR. The Development Team consisted of 14 people: a technical architect, a UI designer, a business analyst, seven developers, three testers, and a technical writer. Eleven of the 14 members of the Development Team had no prior work experience with agile/Scrum and no related training—except for what was provided by the VAR. Of the 16 people on the Scrum Team, 15 were local (at an office in the greater Los Angeles area), and one—the Product Owner—was based out of her office in Paris, France.

Of the 16 people on the Scrum Team, two—the Scrum Master and the Product Owner—were full time on the project. All of the others were allocated about 50% on the project.

I was given a copy of the training binder left by the VAR. I was told that the contents—about 500 pages—reflected the totality of the training and reference material. The training session, led by the VAR, was attended by the Scrum Master and all of the Development Team members. The EVP attended portions. The Product Owner attended parts, listening by phone. The first page in the binder covered the Agile Manifesto, the second page was a two-column table which compared and contrasted waterfall and Scrum (e.g., waterfall freezes scope, Scrum freezes schedule), and the third page showed success rates of waterfall vs. Scrum (e.g., 29% of waterfall projects fail vs. 9% of agile projects). The remaining pages provided information about the VAR company and detailed instructions on how to use their software product. That constituted the training.

The Product Owner, Scrum Master, and Development Team reported that they followed the training and instructions provided by the VAR. I won't delineate the then-current roles further or describe all of the events and artifacts. However, some examples follow.

The Product Owner created a product vision statement and stored it in the software, but nobody else remembered seeing it.

The Product Owner created a product backlog in the software, but nobody else claimed to have seen it.

The Scrum Master facilitated a Sprint Planning event where the Development Team estimated work in terms of complexity, and the results were recorded in the software tool. It was reported that—due to the time difference—

the Product Owner did not attend Sprint Planning meetings.

It was communicated that during Sprint execution, the Scrum Master would ask the Development Team if they had any notable progress; and only if the answer was yes, there was a Daily Scrum. As a result, the Daily Scrum event occurred around once or twice a week. When the meeting took place, the Scrum Master did a quick interview with each member of the Development Team and noted the results in the software tool.

I was told that the team followed the recommendation of the VAR for the sprint duration of four weeks.

The Sprint Reviews were attended by the Scrum Master, the Product Owner (remotely), all of the Development Team members, and the EVP. However, the other stakeholders did not attend the events. On average, about half of the work planned and committed to the sprint was "done." Both "done" and not-"done" items were demonstrated at the Sprint Review.

The Scrum Master reported that the team did conduct a Retrospective event at the end of each sprint and that the results were saved in the software tool. When I reviewed the information, I saw comments such as "we worked very hard" under the what went well category. Everything under the "what didn't go so well/opportunities for improvements" category were ideas for enhancement requests for the software tool. I was informed that the VAR instructed staff to convert everything that didn't go well into a suggestion for a future general release of the software or a request for a custom enhancement of the software.

None of the work from any of the sprints was released.

Part Four

The Options

After gaining a broader and deeper understanding of the organization including their Scrum implementation, I met with the EVP, and we discussed next steps. I presented three options:

1. No change,
2. Revert to the earlier waterfall-only model, or
3. "We can try different things" (aligned with the value of openness) with the objective of improving their agile implementation and achieving their goals.

I said, "we can try different things" because while I believed that changes would likely result in improvements, success could not be guaranteed. I also said, "we can try different things" because any meaningful change would require the cooperation and collaboration of many people. The EVP decided on the third option: trying different things.

The doing of "different things" started with training. I first met with the EVP. I then met with the Scrum Master in several one-on-one meetings. And since he was committed to education and improvement, he, later on, completed the Certified ScrumMaster (CSM) training and certification. The Product Owner was unable to attend the office in person for one-on-one training, but we communicated by phone and Skype. The EVP soon decided that the Product Owner needed to be co-located with the Scrum Team. He found a new Product Owner within the organization. I met with the new Product Owner in multiple one-on-one sessions. And—similar to what occurred with the Scrum Master—since the new Product Owner was committed to education and improvement, he subsequently completed the Certified Scrum Product Owner (CSPO) training and certification. I delivered training to the Development Team as a group. It included an overview and more in-depth coverage of certain topics such as pair programming and technical debt. Later on, some Development Team members completed the Certified Scrum Developer (CSD) training and certification. Stakeholder training follows next.

I delivered a one-hour overview of Scrum to the executive stakeholders in a group session. The attendees asked questions and made comments throughout the meeting. One stakeholder suggested that we do what the American Management Association says is best for Scrum. Then other attendees mentioned additional potential sources for information on agile. I explained that different organizations may have their own perspective on what works well for agile/Scrum, and that one way to go (I mentioned this in part to continue their engagement, involvement, and buy-in) is to look to the Scrum Alliance, a leading authority on the subject, but also see if others have ideas that are aligned with the authority and also fit the desired future state of the organization. I committed to doing the research, and the stakeholders thanked me in advance.

I already had a library of 76 items on agile/Scrum—consisting of material from the Scrum Alliance, Project Management Institute, Deloitte Touche Tohmatsu, Gartner, KPMG, Harvard Business Review, IEEE, MIT, Forbes, and many others. I expanded it to include sources mentioned by stakeholders during the training session, and I diligently reviewed all of the content. The diverse sources identified several values and practices as being central in successful Scrum implementations, and such factors were typically consistent with guidelines from the Scrum Alliance. I then met with each of the stakeholders individually and presented them with a summary of information from the Scrum Alliance and others.

I facilitated a follow-up group meeting with the executive stakeholders. Information from the one-hour training and the diverse sources was summarized. The stakeholders concluded that the central problem with the then-current implementation was that people were not following good practices. They then discussed, voted, and identified what they viewed as the high-level top 10 success factors for a Scrum implementation at the organization based on the previously presented information. In no particular order, the top 10 were:

1. Support from management;
2. Each member of the Scrum Team (Scrum Master, Product Owner, and Development Team) is 100% allocated to the project;
3. There is an Agile Coach, Agile Project Management Office or Agile Center of Excellence;
4. Satisfaction is a crucial metric;

5. The Scrum Team has no more than 11 people;
6. The Scrum Team is co-located;
7. There are consistent practices and processes;
8. There is a digital wallboard or other information radiator;
9. There are frequent and high-quality interactions; and
10. There is continuous improvement/inspect and adapt.

Some of the items overlap/are not mutually exclusive, and the items are not exhaustive. I thanked the executive stakeholders for their support, and I told them that the 10 factors are built into the go-forward plan. The stakeholders expressed their appreciation.

The EVP and I conducted a mini-retrospective on the training sessions. The EVP was enthusiastic about what we've done so far, and he said that morale had improved. He authorized company-paid CSM, CSPO, and CSD trainings and certifications for staff (mentioned earlier). The EVP said we could advance to the next stage: piloting changes. I asked if he was open to terminating the use of the software tool introduced by the VAR. He initially said that so much time and money had been invested in it that it would be hard to justify doing so. I said it differently: "We can try different things" could mean putting the software tool on vacation for a period. He agreed.

Part Five

The Pilot – Vision, Roadmap and Release Plan, and Product Backlog

We advanced to the pilot—which included doing many things differently. Examples follow.

The Product Owner and I discussed techniques on developing a product vision statement. He opted to use the template attributed to Geoffrey Moore. The Product Owner created a draft of the vision and sent it with a request for feedback to the stakeholders. After receiving feedback, the Product Owner revised and finalized the product vision. He wrote the statement on oversize paper and posted it in a prominent location where the Scrum Team and stakeholders could easily see it.

The Product Owner and I then discussed techniques on creating and maintaining a product roadmap/release plan. He opted for a simple table with four rows and four columns. The rows included:

- Name (the title of the product or major release),
- Goal (the reason for creating it),
- Features (a high-level list of features), and
- Estimated number of sprints.

For the columns, there was one for each quarter of the year. Similar to what was done with the product vision, the Product Owner sent the plan with a request for feedback to the stakeholders. After receiving feedback, the Product

Owner revised and the plan, wrote it on oversize paper, and posted it next to the product vision statement.

Aspects of team formation were covered already. The new Scrum team totaled 11 people. All were co-located and 100% allocated to the project.

The Product Owner and I discussed techniques on developing and maintaining the product backlog. He opted to employ a simple table format with six columns:

- ID#,
- User story/description,
- Category (he decided on four types: feature, bug, technical debt, and other),
- Story point estimate for complexity,
- Priority based on business value, and
- Status.

To help with user stories, the Product Owner often referenced the INVEST (Independent, Negotiable, Valuable, Estimable, Small, and Testable) model developed by Bill Wake. For priority, the Product Owner initially used the MoSCoW method (Must have, Should have, Could have, or Won't have). Later, the Product Owner found the business value/risk method (where each item is rated as high or low in two dimensions—business value and risk, ...) to be the most beneficial, and the Product Owner continues to use that approach today. In a session facilitated by the Scrum Master, the Product Owner presented the user stories to the Development Team, and participants provided story points (using physical cards for the exercise) for estimates of complexity of each item. Later, the team tried t-shirt sizing—S, M, L, and XL designations—for estimates of complexity, but they decided to return to story points.

Previously, sprints were four weeks in length. Now the team was using the shorter duration of two weeks. A key benefit was that the Scrum value of focus was improved.

(This page is intentionally blank.)

Part Six

The Pilot (Continued) – Sprint Planning and Sprint Execution

The Product Owner, Scrum Master and I discussed sprint planning techniques. The Scrum Master decided that the meeting event would be handled via two separate sessions—part 1 (what will be committed to for the upcoming sprint) and part 2 (how to accomplish the work identified in part 1).

For sprint planning part 1, the timebox (not to exceed duration) for the meeting was calculated as:

- 2
- multiplied by the number of weeks in the upcoming sprint (2 in this case),
- which equaled 4 hours for the event.

The Scrum Master made the following information visible during the event: start and end dates for the sprint, (after a sprint was completed) the results of the last sprint review event, and (after a sprint was completed) the results from the last sprint retrospective event. The Product Owner reminded the Development Team about the product vision statement, and the Product Owner shared the sprint goal (such as "implement shopping cart functionality ..."). The Development Team determined their capacity in work hours for the upcoming sprint. It was calculated as:

- the number of people in the Development Team
- multiplied by the number of project productive hours (which excluded time outside the sprint such as company meetings, trainings, vacation time, etc.) per workday
- multiplied by number of workdays in the sprint.

(Estimation via story points, and prioritization by the Product Owner were already taken care of.)

For each item in the product backlog, participants discussed the user stories/requirements including acceptance criteria, assumptions, dependencies, risks, and anything else requiring a conversation to get a good understanding of the item. The Development Team then committed to the entries which they thought could be completed in the upcoming sprint. The technique they employed involved asking "Can we do this first item in the product backlog?" If the answer was yes, they selected it and proceeded to the next item and continued until the team believed that no more work could be done in the sprint. After the Development Team had worked together and had data on actual velocity (the number of story points completed in a sprint), they also considered that historical metric—comparing it with story points for items in the sprint. The Product Owner updated the product backlog, identifying the items committed by the Development Team to be done for the upcoming sprint.

For sprint planning part 2, the timebox for the meeting was calculated as:

- 2
- multiplied by the number of weeks in the upcoming sprint (2 in this case),
- which equaled 4 hours for the event.

The Scrum team created the sprint backlog. It had the following columns:

- ID#,
- Description,
- Story points,
- Task information (meetings, designs, coding, code review, testing, etc.),
- Estimation in hours (they adopted the practice that if effort is greater than eight hours, split the task into smaller ones),
- Owner (where members of the Development Team self-assign tasks),
- Status, and
- Hours of work remaining.

During the meeting, the Scrum team compared the total estimated work hours for the sprint with the Development Team's capacity (mentioned under the part 1 meeting) for the sprint. If the Development Team believed that the sprint backlog contained too much work to be done during the sprint, they collaborated with the Product Owner to remove one or more items. If the Development Team believed they could handle more work during the sprint, they worked with the Product Owner to move one or more of the high priority items from the product backlog to the sprint backlog.

The Product Owner, Scrum Master and I discussed sprint execution. Select examples of what was decided and done are highlighted next.

The Development Team set up a task board (also known as a Scrum board) to reflect the work in the current iteration. They went with a simple format. The board depicted work in rows and columns where rows included work items, and columns reflected status (To Do, Doing, and Done). Work was addressed from top (highest priority) to bottom, and work migrated from left to right on the task board as it progressed. The task board is also covered in the daily Scrum meeting.

The Scrum Master decided to use a sprint burndown chart to track and communicate progress during the sprint. He set it up and updated it each workday, usually immediately after the daily Scrum meeting.

The Scrum Master created an impediment backlog to capture things preventing the team from progressing or improving. This backlog was updated daily, typically immediately after the daily Scrum meeting.

At the daily Scrum meeting event, the Development Team shared status, plans, and any impediments. Before this pilot with changes, the team was not conducting the daily Scrum (or updating the task board, burndown chart, and impediments backlog) consistently. Under the pilot (and subsequently), the Development Team met for up to 15 minutes (timebox) each workday, and it was conducted at the same time (10:00 a.m.) each day. At this daily stand-up session, the Development Team and the Scrum Master met where the task board, sprint burndown chart, and impediment backlog were posted. Each Development Team member described what he/she worked on since the last Scrum meeting, and he/she updated the task board. Next, the same Development Team member explained what he/she would work on that day, and he/she updated the task board. Lastly, the same Development Team member reported any impediments. (The Scrum Master recorded any issues in the impediments backlog. If a discussion was required, it took place immediately after the daily Scrum. The Scrum Master helped resolve impediments.) The steps were repeated for other members of the Development Team.

The Scrum Team built an increment of functionality during every sprint, and the increment was potentially shippable because the Product Owner might decide for it to be implemented at the end of the sprint. Said differently, potentially shippable is defined by a state of confidence or readiness, and shipping is a business decision. Commencing with the pilot, the organization started releasing products as often as every sprint (two weeks).

(This page is intentionally blank.)

Part Seven

The Pilot (Continued) – Sprint Review and Sprint Retrospective

The Scrum team—Product Owner, Scrum Master, and the Development Team—and I discussed techniques for the sprint review. The event is sometimes referred to as the sprint demo. The timebox (not to exceed duration) in hours for the meeting was calculated as:

- 1
- multiplied by the number of weeks in the upcoming sprint (2 in this case),
- which equaled 2 hours for the event.

Participants included the Product Owner, Scrum Master, Development Team, and stakeholders. All of the stakeholders (the executives mentioned in earlier parts of the story) were invited to and attended the sprint review. At the session, the Product Owner welcomed attendees and communicated the agenda. He pointed out the sprint goal, which was displayed on the wall in the meeting room. Next, the Development Team listed the work that was committed to the sprint. They then listed the work that was completed and the work that was not completed. For each completed story/feature, the Development Team demonstrated the "done" working functionality and answered questions. Stakeholders were invited to interact with the "done" working functionality and they did so. Then the entire group reviewed the product backlog and collaborated on what to do next. The Product Owner incorporated feedback into the Product Backlog. It involved adding new items to the backlog and/or re-prioritizing existing items. After the sprint review, the Scrum Master incorporated any feedback related to problems into the impediments backlog.

The Scrum Master and I discussed techniques for the sprint retrospective event. The timebox in hours for the meeting was calculated as:

- 0.75
- multiplied by the number of weeks in the upcoming sprint (2 in this case),
- which equaled 1.50 hours for the event.

At the sprint retrospective, the Development Team identified what went well, what didn't go well, and improvements to be implemented in the upcoming sprint—an example of inspection and adaptation. For improvements to be implemented, each member of the Development Team wrote their top one, two or three suggestions on sticky notes. Each idea got its own sticky note. Then the Development Team grouped the sticky notes into categories/themes, and they discussed the items and voted to determine the top one (initially limited to one, later it was raised to two) for adaptation and improvement. The Development Team committed to the change, and the Scrum Master recorded the information.

Part Eight

The Success

By adopting an agile mindset and providing improved engagement, collaboration, transparency, and adapt-ability via Scrum's values, roles, events, and artifacts, the results were excellent. After one sprint, satisfaction ratings for the Development Team and stakeholders were higher than the target. After three sprints, the output of the Scrum Team became consistent and predictable, satisfaction increased even further, and all of the seven goals mentioned earlier were achieved. Here's an overview:

Goal 1

- Source: This goal was identified by the EVP.
- Desired outcome: Develop and deliver products and services faster than 6-12 months.
- Result: The goal was achieved (with the agile transformation, delivery occurred as often as every two weeks).

Goal 2

- Source: This goal was identified by the EVP.
- Desired outcome: Improve the satisfaction of the Scrum Team.
- Result: The goal was achieved.

Goal 3

- Source: This goal was identified by the EVP.
- Desired outcome: Improve the satisfaction of the stakeholders.
- Result: The goal was achieved.

Goal 4

- Source: This goal was identified by the Scrum Team (Scrum Master, Product Owner, and the Development Team).
- Desired outcome: Meet or exceed the expectations of management.
- Result: The goal was achieved.

Goal 5

- Source: This goal was identified by the Scrum Team.
- Desired outcome: Deliver valuable products/ services.
- Result: The goal was achieved.

Goal 6

- Source: This goal was identified by the stakeholders.
- Desired outcome: For stakeholders or their representatives to be more involved in requirements.
- Result: The goal was achieved.

Goal 7

- Source: This goal was identified by the stake-holders.
- Desired outcome: Get useable products delivered more frequently than 6-12 months.
- Result: The goal was achieved (with the agile transformation, delivery occurred as often as every two weeks).

The EVP, Scrum Team, and stakeholders declared the pilot a success; and they made this implementation of Scrum the preferred approach for the development and delivery of products and services going forward.

In addition to achieving the goals summarized above, the agile transformation also supported the organization's efforts to drive innovation and be more competitive. As the business became more prosperous and grew, so did the number of Scrum teams. Originally numbering one, the division presently has six Scrum teams.

This concludes the brief, behind-the-scenes account of how a global entertainment company experienced a successful agile transformation and dramatically improved the way it works and delivers business value. I hope that you enjoyed it.

(This page is intentionally blank.)

If You Enjoyed This Book ...

... please leave a review on Amazon.

... you can learn even more from *Agile Scrum: Your Quick Start Guide with Step-by-Step Instructions*. It helps technical and non-technical teams develop and deliver products in short cycles with rapid adaptation to change, fast time-to-market, and continuous improvement—which supports innovation and drives competitive advantage. Here's what some reviewers have said about it:

"The book highlights the versatility of Scrum beautifully."

— Literary Titan

"A must-have for a project manager wanting to introduce Scrum to the organization."

— PM World Journal

"A superbly written and presented guide to team-based project management that is applicable across a broad range of businesses from consumer products to high-tech."

— IndieBRAG

"*Agile Scrum: Your Quick Start Guide with Step-by-Step Instructions* is an all-inclusive instruction guide that is impressively 'user-friendly' in tone, content, clarity, organization, and presentation."

— Midwest Book Review

"A clear and authoritative roadmap for successful implementation, *Agile Scrum: Your Quick Start Guide with Step-by-Step Instructions* is strongly recommended."

— BookViral

— Readers' Favorite

— Amazon.com Hall of Fame and Top 100 Reviewer

Agile Scrum: Your Quick Start Guide with Step-by-Step Instructions has garnered 17 first place wins. Credit is shared with Chris Hare and Colin Giffen, the technical editors on the publication.

Scott and his book have been featured in Yahoo Finance, the Boston Herald, NBC WRAL, Computer Weekly, the Dallas Business Journal, the PM World Journal, BookLife by

Publisher's Weekly, Learning Solutions, Innovation Management, and additional media publications.

A trailer, high-resolution images, reviews, a detailed list of awards, and more are available in the digital media kit at AgileScrumGuide.com.

Agile Scrum: Your Quick Start Guide with Step-by-Step Instructions is available in digital and print formats. The ebook is for sale in Australia, Brazil, Canada, France, Germany, India, Italy, Japan, Mexico, the Netherlands, Spain, the United Kingdom, and the United States. The paperback is available at Amazon, Barnes & Noble, Strand Books, Harvard Book Store, Books-a-Million, The Booksmith, Hudson Booksellers, Savoy Bookshop & Café, Compass Books at SFO/Books Inc., Books & Books - Miami, University Press Books - Berkeley, and other retailers, distributors, and partners around the globe.

(This page is intentionally blank.)

About the Author

Scott M. Graffius is a technology leader, project management expert, consultant, international speaker, and award-winning author.

Scott is a Principal Consultant and the CEO of Exceptional PPM and PMO Solutions™, a professional services firm, where he partners with client companies to help them strengthen their project management capabilities and realize their strategic objectives and business initiatives. The consultancy provides advisory, training, and facilitative consulting services related to project, program, portfolio, and PMO management. The firm's expertise spans agile, traditional waterfall, and hybrid approaches. While every engagement is unique, business outcomes typically include getting more projects done, faster delivery time, improved on-budget performance, better management of risks, improved customer and stakeholder satisfaction, and more consistent realization of business results. Exceptional PPM and PMO Solutions™ confidently backs its services with a Delighted Client Guarantee™. Details are available at Exceptional-PMO.com.

A fantastic agile transformation experience and result with a client in the entertainment industry was the inspiration for Scott's first book, *Agile Scrum: Your Quick Start Guide with Step-by-Step Instructions* (ISBN-13: 9781533370242). It helps technical and non-technical teams develop and deliver products in short cycles with rapid adaptation to change, fast time-to-market, and continuous improvement—which supports innovation and drives competitive advantage. The book has garnered 17 first place awards from national and international competitions. Scott and *Agile Scrum: Your Quick Start Guide with Step-by-Step Instructions* have been featured in Yahoo Finance, the Boston Herald, NBC WRAL, Computer Weekly, the Dallas Business Journal, the PM World Journal, BookLife by Publisher's Weekly, Learning Solutions, Innovation Management, and additional media publications. A trailer, high-resolution images, reviews, a detailed list of awards, and more are available in the digital media kit at AgileScrumGuide.com.

Scott's second title, *Agile Transformation: A Brief Story of How an Entertainment Company Developed New Capabilities and Unlocked Business Agility to Thrive in an Era of Rapid Change* (ASIN: B07R9LJLPJ), was published in April 2019. BookAuthority named it one of the best new books on Scrum.

Scott is a former vice president of project management with a publicly traded provider of diverse consumer products and services over the Internet. Before that, he ran and supervised the delivery of projects and programs in public and private organizations with businesses ranging from e-commerce to advanced technology products and services, retail, manufacturing, entertainment, and more. He has experience with consumer, business, reseller, government, and international markets, as well as experience spanning 20 countries.

Scott has a bachelor's degree in Psychology with a focus in Human Factors. He holds seven professional certifications:

Certified Scrum Professional - ScrumMaster (CSP-SM), Certified Scrum Professional - Product Owner (CSP-PO), Certified ScrumMaster (CSM), Certified Scrum Product Owner (CSPO), Project Management Professional (PMP), Lean Six Sigma Green Belt (LSSGB), and IT Service Management Foundation (ITIL). He is a member of the Scrum Alliance, the Project Management Institute, and the Institute of Electrical and Electronics Engineers (IEEE). Within the IEEE, Scott is a member of the Computer Society, the Consumer Electronics Society, the Broadcast Technology Society, the Internet of Things (IoT) Community, and the Consultants Network.

He has been actively involved with the Project Management Institute (PMI) in the development of professional standards. He was a member of the team which produced the *Practice Standard for Work Breakdown Structures—Second Edition*. Scott was a contributor and reviewer of *A Guide to the Project Management Body of Knowledge—Sixth Edition* and *The Standard for Program Management—Fourth Edition.* He was also a subject matter expert reviewer of content for the PMI EMEA Congress 2019.

Scott regularly speaks at conferences and other events in the United States and internationally where he delights audiences with presentations on agile, traditional, and hybrid project, program, portfolio, and PMO management. He integrates content on professional standards, best practices, and his first-hand experience with successful implementations. Scott uses everyday language and vibrant custom visuals to make complex topics clear and understandable, and he provides audiences with practical information they can use. His upcoming and past speaking engagements are listed at ScottGraffius.com.

Scott resides in Los Angeles, California.

Connect on Social

You're invited to connect with ...

... Scott on Facebook, Twitter, LinkedIn, and Amazon. You'll find links to those accounts at ScottGraffius.com.

... Scott's business, Exceptional PPM and PMO Solutions™, on Twitter, Facebook, and LinkedIn. You'll find links to those accounts at Exceptional-PMO.com.

... *Agile Transformation: A Brief Story of How an Entertainment Company Developed New Capabilities and Unlocked Business Agility to Thrive in an Era of Rapid Change* on Twitter, Facebook, Instagram, LinkedIn, and Vimeo. You'll find links to those accounts at bit.ly/at-site.

... Scott's other book, *Agile Scrum: Your Quick Start Guide with Step-by-Step Instructions*, on Twitter, Facebook, Instagram, Pinterest, LinkedIn, and Vimeo. You'll find links to those accounts at AgileScrumGuide.com.

(This page is intentionally blank.)

Glossary

Terms mentioned in the story are listed in this glossary. For additional value, it also includes terms on agile and Scrum *not* in the story.

Acceptance criteria: Details that indicate the scope of a user story and help the team determine "done"-ness.

Agile: An iterative and incremental development method. It promotes evolutionary development and is designed to support rapid and flexible responses to changing requirements.

Agile coach: The individual is an agile expert who provides guidance for new agile implementations as well as existing agile teams. The agile coach is experienced in employing agile techniques in different environments and has successfully run diverse agile projects. The individual builds and maintains relationships with everyone involved, coaches individuals, trains groups, and facilitates interactive workshops. The agile coach is typically from outside the organization, and the role may be temporary or permanent.

Agile Manifesto: A formal proclamation of four key values and twelve principles to guide an iterative and people-centric approach to development.

Agile principles: The twelve principles that underpin the Agile Manifesto. See **Agile Manifesto**.

Artifacts: Documents or visual depictions of work items, progress, features, code base, etc.

Automated integration testing: Where individual software modules are automatically combined and tested as a group.

Backlog: See **Product backlog** and **Sprint backlog**.

Blockers: See **Impediment**.

Burndown: See **Sprint burndown chart** and **Product burndown chart**.

Burn rate: The rate at which hours allocated to a project are being used.

Cadence: A regular, predictable rhythm. Sprints of consistent duration (two weeks, for example) establish a cadence for development effort.

Cancellation: See **Termination**.

Ceremonies: See **Events**.

Chart: See **Sprint burndown chart** and **Product burndown chart**.

Continuous integration: Where individual software modules are combined and tested with existing software as soon as they are produced.

Core roles: There are three core roles—Product Owner, Scrum Master, and Development Team. The core roles collectively constitute the Scrum team.

Cross-functional team: A group of people with different expertise working towards a common goal.

Daily Scrum: Each day during a sprint, the team holds a daily Scrum meeting. It lasts no more than fifteen minutes, and it provides a quick update on progress. Each member of the Development Team reports on tasks finished the prior workday, work to be done today, and any impediments blocking progress. This event is sometimes referred to as the daily stand-up or the daily sync.

Daily Stand-up: See **Daily Scrum**.

Definition of Done (DoD): The exit-criteria to determine whether a product backlog item is complete. It provides precision about when work is complete. The DoD typically involves the following: Code complete, unit tests written and executed, integration tested, performance tested, documented, and accepted by Product Owner. The DoD may vary from one Scrum team to another, but it must be consistent within a team.

Demo: See **Sprint review**.

Development Team: A cross-functional group of people responsible for delivering potentially shippable increments of a product at the end of every sprint. The team is comprised of between three and nine (a previous guideline was between five and nine) people—developers, testers, business analysts, etc. Their responsibilities include being self-organizing, negotiating commitments with the Product Owner, collaborating with anyone necessary to get the job done, delivering the product, alerting the Scrum Master of any impediments, presenting the product at the sprint review (demo), and inspecting and adapting the process.

DoD: See **Definition of Done**.

Done: See **Definition of Done**.

Estimated work remaining: The hours that a team member estimates that remains to be worked on for a task. This estimate is updated each day that the task is worked on.

Events: There are four key events (meetings)—Sprint planning, Daily Scrum, Sprint review, and Sprint retrospective. Events were previously known as ceremonies. Also see: **Sprint planning, Daily Scrum, Sprint review**, and **Sprint retrospective**.

Fibonacci sequence: Attributed to Italian mathematician Leonardo Pisano, the Fibonacci sequence is a series of numbers where a value is found by adding up the two numbers before it. Starting with 0 and 1, the sequence goes 0, 1, 1, 2, 3, 5, 8, 13, 21, 34, etc. In Agile Scrum, story points are used to rate the difficulty to implement a story (work). Most Scrum teams use the Fibonacci sequence or a variation as an estimating technique.

Functionality: The behaviors that a computer system is designed to achieve.

Grooming: See **Product backlog refinement**.

Impediment: Issues or anything else that prevents a team member from performing work as efficiently as possible. Impediments may also be known as blockers, issues or problems.

Increment: Potentially shippable completed work that is the outcome of a Sprint.

Information radiator: A large graphical representation of key project information, updated regularly. The information is displayed near the team's workspace, accessible to stakeholders. The information may be presented on flip chart paper or a whiteboard or displayed on a monitor. Information radiators are sometimes called wallboards or electronic wallboards.

INVEST: Introduced by Bill Wake, INVEST is an acronym for a model for developing a well-written user story. INVEST stands for Independent, Negotiable, Valuable, Estimable, Small, and Testable. Also see: **User story**.

Issue: See **Impediment**.

Iteration: A short time period in which a team is focused on delivering an increment of a product that is useable. Also see: **Sprint**.

Meetings: See **Events**.

Minimum viable product: The MVP has just those features (functional, reliable and usable) considered sufficient for it to be of value to customers, and allow for it to be shipped or sold to early adopters. Customer feedback will inform future development of the product.

MoSCoW: MoSCoW is a popular acronym that represents a method of ranking stories (requirements, new features, bug fixes, etc.). With MoSCoW, each item is sorted into one of four categories: Must have, Should have, Could have, or Won't have. Also see: **User story**.

Pair programming: An agile software development technique in which two programmers collaborate on the same task on a single computer. The person who controls the mouse and keyboard is called the "driver." The other person, who sits beside the driver and helps ensure that the solution is implemented in an effective and efficient manner, is called the "navigator." The members of the pair can switch with other members of the team to take on different tasks. The key objective is to produce high-quality code. Benefits typically include no separate code reviews required, better code quality, effective communication, better coding practice adherence, a more effective inclusion of new team members, and greater knowledge sharing among team members.

Potentially shippable product: An increment of work that is complete per the Definition of Done and is capable of being released.

Problem: See **Impediment**.

Product backlog grooming: See **Product backlog refinement**.

Product goal: See **Product vision statement**.

Product increment: The increment or potentially shippable increment (PSI) is the sum of all the product backlog items completed during a sprint and all previous sprints. At the end of a sprint, the increment must be complete, according to the Scrum team's Definition of Done, and in a usable condition regardless of whether the Product Owner decides to release it.

Product Owner: The person responsible for maintaining the product backlog by representing the interests of the stakeholders and ensuring the value of the work done by the Development Team. More specifically, the Product Owner is responsible for representing the voice of customers, communicating with stakeholders to ensure their interests are represented, managing stakeholder expectations, establishing and achieving the product vision, defining releases, defining sprint goals, managing the return on investment, creating and maintaining the product backlog, authoring and prioritizing user stories based on business value, outlining acceptance criteria, attending the sprint reviews and planning sessions, and continuously reprioritizing the product backlog.

Product backlog: A list of functional and non-functional requirements, usually expressed as user stories. Entries are prioritized based on business value.

Product backlog refinement: An ongoing process of adding detail and estimates, as well as re-ordering the backlog items. It is sometimes referred to as **Product backlog grooming**.

Product burndown chart: This depicts the points of all user stories from the product backlog. It shows the story points for completed work in each sprint, illustrating the completion of requirements over time. Also see: **Sprint burndown chart** and **Information radiator**.

Product vision statement: The product vision is the overall objective. It's created by the Product Owner, and it describes in about two sentences the product's purpose, who it is for, how it will create value, and the benefits. The vision often encompasses multiple releases. The product vision is sometimes called the **Product goal**.

Refactoring: The process of improving code—by clean-up, simplification, etc.—to make it easier to maintain and expand in the future. Refactoring is a necessary step to keep the cost of changes low.

Refinement: See **Product backlog refinement**.

Release: The transition of the final product into routine use by the end user.

Retrospective: See **Sprint Retrospective**.

Scrum: The most popular agile development and delivery framework. It encompasses a powerful set of principles and practices that help teams deliver products in short cycles, which promotes speedy feedback, rapid adaptation to change, faster delivery time, and continuous improvement.

Scrum components: Scrum's roles, events, artifacts, and the rules that bind them together.

Scrum Master: The individual who ensures the team adheres to Scrum practices, values, and rules. His/her responsibilities include serving as single point of contact for the project, facilitating and enforcing the Scrum process, coaching the team on Scrum values and practices, facilitating the daily Scrum (stand-up meeting), ensuring the team is fully functional and productive, helping the team remove impediments, capturing data, keeping Scrum artifacts (charts, etc.) current and visible, enforcing time-boxes, promoting improved engineering practices, shielding the team from external interferences and distractions, and conducting the sprint retrospective at the end of a sprint.

Scrum team: The Scrum team is made up of the Product Owner, Scrum Master, and Development Team. It totals between five and eleven (between seven and eleven in a prior guideline) people.

Scrum values: A set of fundamental values and qualities underpinning the Scrum framework: focus, courage, openness, commitment, and respect.

Self-organizing: The principle that teams autonomously organize their work. Self-organization happens within boundaries and against given goals. Self-organizing teams determine how to best accomplish their work, rather than being directed by others outside the team.

Show and tell: See **Sprint review**.

Spike: A time-boxed period used to research a concept or create a prototype. Spikes typically occur between sprints. Unlike sprints, spikes may or may not deliver shippable and valuable functionality.

Sprint: A period of one, two, three or four weeks (two is most common) during which the team will work on a set of backlog items that were committed to be completed.

Sprint backlog: A prioritized list of tasks to be completed during the sprint. Also see: **MoSCoW**.

Sprint burndown chart: Shows the work remaining in the sprint backlog. It is refreshed before the next daily Scrum meeting. Also see: **Product burndown chart** and **Information radiator**.

Sprint goal: The purpose of a Sprint. It's often expressed as a proposed solution to a business problem.

Sprint planning: A time-boxed event (meeting), set to run two hours for each week of the sprint. It occurs before the sprint, and there are two parts to the event. During the first half, the sprint team (Product Owner, Scrum Master, Development Team) agree on what product backlog items to consider for the sprint. During the second half, the Development Team decomposes the work required to deliver the backlog items, resulting in the sprint backlog.

Sprint retrospective: At the end of a sprint, the team holds two meetings—the sprint review and the sprint retrospective. At the sprint retrospective, the team reflects on the past sprint, and identifies and prioritizes improvements for the next sprint. The event is time-boxed to run 45 minutes for each week of the sprint and is facilitated by the Scrum Master. Also see: **Sprint review**.

Sprint review: At the end of a sprint, the team holds two events (meetings)—the sprint review and the sprint retrospective. At the sprint review, the team reviews the work that was completed and the planned work that was not completed, and demonstrates the completed work. The event is time-boxed to run one hour for each week of the sprint. Also see: **Sprint retrospective**.

Sprint tasks: Work items added to the sprint backlog at the beginning of a sprint and broken down into hours. It is recommended that each task not exceed six hours (or one work day).

Stakeholder: Someone with an interest in the outcome of a project, either because he or she has funded it, will use it or will be affected by it.

Stand-up meeting: See **Daily Scrum**.

Story points: Used to rate the difficulty (related to complexity, unknowns, etc.) to implement a story. The Fibonacci sequence (0, 1, 2, 3, 5, 8, 13, 21, ...) and t-shirt sizes (small, medium, large, extra-large) are common scales. Also see: **User story** and **Fibonacci sequence**.

Task board: A notice board that shows the progress of each task. The most basic states for status are: "To Do", "Doing," and "Done." The task board can be physical or shared via a software solution.

Tasks: See **Sprint tasks**.

Team: See **Scrum team** and **Development Team**.

Team capacity, estimated: This is calculated as the number of sprint Development Team members, multiplied by the number of project productive hours per workday, multiplied by the number of work days in the sprint. Project productive hours should exclude time outside the sprint, such as company meetings, trainings, vacation time, etc.

Technical debt: The cumulative total of poor design and coding. It may have one or more causes such as time pressures, overly complex technical design, lack of alignment to standards, suboptimal code, delayed refactoring, insufficient testing or inadequate documentation. The consequence of technical debt is that more time is needed later on in the project to resolve issues. It can be avoided or minimized by not taking shortcuts, using simple designs, and refactoring continuously. When there's technical debt, the team should make the items visible by registering entries in the product backlog, where the items will be evaluated and prioritized for resolution.

Termination, Abnormal: The Product Owner can cancel a sprint if necessary. Changes in external market conditions may be a reason, for example. If a sprint is abnormally terminated, the next step is to conduct a new sprint planning session, where the reason for the termination is reviewed.

Testing: In Agile Scrum, the approach includes testing performed early and often, and close cooperation with developers and customers/users. It often involves unit testing, application program interface and service testing, acceptance testing, system testing, regression testing (to detect any side effects from changes), and user acceptance testing (UAT). Tests should be automated as much as possible.

Time-box: Setting a duration for an activity and not exceeding it. For example, a daily Scrum meeting is time-boxed to 15 minutes and ends no later than that time.

User story: Each user story represents a portion of business value that a team can deliver in an iteration. They act as requirements and are written in plain language. The format of a user story is: "As a <Role>, I want <goal> so that I can <reason>." Example: "As a customer, I want shopping cart functionality so that I can buy items online." User stories are captured in the product backlog. Also see: **INVEST**, **Story points** and **Fibonacci sequence**.

Velocity, Actual: The sum of the team's delivery of completed work from the product backlog. It is usually measured in story points. Example: A team delivers story "A," which had 4 points, story "B" with 7 points, and story "C" with 10 points; thus, the velocity is 21.

Velocity, Planned: This is the expected velocity for the team based on historical data. The team's actual velocity history (the average, for example) is used for planning future sprints/iterations.

Wallboard: See **Information radiator**.

Bibliography

Graffius, Scott M. (2016). *Agile Scrum: Your Quick Start Guide with Step-by-Step Instructions*. North Charleston, SC: CreateSpace.

(This page is intentionally blank.)

Appendix

Agile Manifesto

In February of 2001, a group of software engineers met and brainstormed ways to improve software development. They developed *The Agile Manifesto.* It says:

> We are uncovering better ways of developing software by doing it and helping others do it. Through this work we have come to value:
>
> - Individuals and interactions over processes and tools
> - Working software over comprehensive documentation
> - Customer collaboration over contract negotiation
> - Responding to change over following a plan
>
> That is, while there is value in the items on the right, we value the items on the left more.

"Items on the left" refers to the content to the left of the word "over." For example, "following a plan" is valuable but "responding to change" is more valuable.

The original manifesto authors included Kent Beck, Alistair Cockburn, James Grenning, Ron Jeffries, Robert C. Martin, Jeff Sutherland, Mike Beedle, Ward Cunningham, Jim

Highsmith, Jon Kern, Steve Mellor, Dave Thomas, Arie van Bennekum, Martin Fowler, Andrew Hunt, Brian Marick, and Ken Schwaber.

Agile Principles

In 2001, a group of software engineers met and brainstormed ways to improve software development. They developed *The Agile Manifesto*. As a follow up to the manifesto, they established the following philosophies.

Twelve Principles of Agile Software

1. Our highest priority is to satisfy the customer through early and continuous delivery of valuable software.
2. Welcome changing requirements, even late in development. Agile processes harness change for the customer's competitive advantage.
3. Deliver working software frequently, from a couple of weeks to a couple of months, with a preference to the shorter timescale.
4. Business people and developers must work together daily throughout the project.
5. Build projects around motivated individuals. Give them the environment and support they need, and trust them to get the job done.
6. The most efficient and effective method of conveying information to and within a Development Team is face-to-face conversation.
7. Working software is the primary measure of progress.
8. Agile processes promote sustainable development. The sponsors, developers, and

users should be able to maintain a constant pace indefinitely.

9. Continuous attention to technical excellence and good design enhances agility.
10. Simplicity – the art of maximizing the amount of work not done – is essential.
11. The best requirements and designs emerge from self-organizing teams.
12. At regular intervals, the team reflects on how to become more effective, then tunes and adjusts its behavior accordingly.